# *Lattice Labyrinth*
# *Tessellations*

## bold art from modest mathematics

## David Mitchell

# tarquin

# Contents

© 2013: David Mitchell
ISBN: 978 1 907550 850
Printed and designed in the UK

tarquin publications
Suite 74, 17 Holywell Hill
St Albans, AL1 1DT, UK
www.tarquingroup.com

Distributed in the USA by Parkwest
www.parkwestpubs.com
www.amazon.com & major retailers

Distributed in Australia by OLM
www.lat-olm.com.au

# Introduction

Welcome to the No-Man's-Land of *Lattice Labyrinth Tessellations*, an exciting, fertile, yet almost unexplored Flatland situated in the gulf between Art and Mathematics. Don't be afraid; the subject is accessible to anyone of secondary-school age and upwards – I'm a long way upwards myself. The astonishing and elegant designs revealed in this very short and almost elementary book are simple to make and beautiful to contemplate, yet are unmentioned in the textbooks and unknown to designers and architects. You can share in the joy of generating decorative patterns that have a simple, fundamental origin and yet may never before have been seen by mortal eyes. Many of the shapes found will suggest human or other living shapes that you may be able to transmute into striking images, following in the footsteps of the celebrated Dutch graphic designer Maurits Escher.

This is a guided workbook that will show you how to construct an unbounded range of potentially intricate patterns like those on the front and back covers. Many of these designs, in fact an infinite number, can be constructed, step-by-step, following straightforward rules. Once you're familiar with the steps and the rules, you may get the bug. You may even join me in the challenging pastime of recreational wrestling with some of the infinitely many more cases for which sufficiently general rules are yet to be found and every drawing is an experiment.

If you look closely, you will see that each pattern on the covers is made up of identical shapes that fit together perfectly like the pieces of a jigsaw puzzle. We say that these shapes *tessellate* and we call the patterns they make when fitted together *tessellations*. Here are some more examples of shapes that tessellate; I call them *supertiles*.

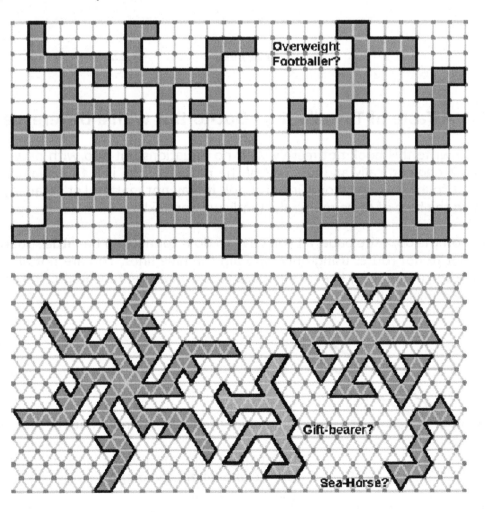

Overweight Footballer?

Gift-bearer?

Sea-Horse?

Notice that all these supertiles are made up either of square or of equilateral triangular tiles and that their shapes, though often convoluted, are subject to stern discipline. Not only do they tessellate, but they are (with one exception) highly symmetrical and their arms, however far flung, branching and tortuous, are never more than one tile wide. These arms intertwine in the tessellation, their tips penetrating to within one tile of the centre of each neighbouring supertile. (The exceptional, non-symmetrical "gift-bearer" supertiles interlock to form a tessellation which is symmetrical as a whole – see the final "advanced" exercise on page 39)

I call the tessellations made up of these supertile jigsaw pieces *Labyrinths* because their intricacy can be baffling to the eye despite their fundamental simplicity (they can be specified using very little information; I think a computer file of one could be very strongly compressed). They are called *Lattice Labyrinths* because they can all be set out on the *Square Lattice* or the *Triangular Lattice*, which are networks of points joined by straight lines; "graph-paper-with-dots". All the supertiles on page 1 are shown set out on one or other of these lattices. Here they are again, unobstructed by supertiles.

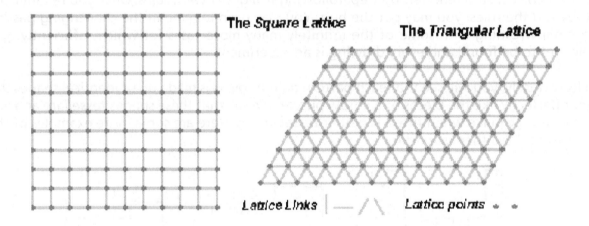

## A Taste of Tessellating - Dot-to-Dot with a Difference

I wish I could write this book not merely with no equations but also with no words. It's much easier to think geometrically and just DRAW the patterns than it is to explain them in words, so if you can't follow ALL the words in this book it may not matter. Let's start by completing a drawing, with the help of some more reluctantly written text.

In the figure on the opposite page, all you have to do is connect up the little square and lozenge symbols by dot-to-dot routes using only (and all) the pale lines. You are not allowed to follow any of the thicker, darker-grey lines of what I call the complementary graph. The enlarged section below shows you what I mean.

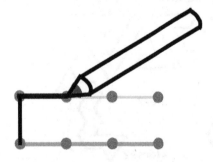

Draw over these pale, thinner lines (lattice links)

Don't draw over the thicker, darker lines (where the "*complementary graph*" is already covering the lattice links).

Your goal is to pencil over all the lighter lines. I suggest you use a pencil in case of mistakes and then ink the lines over when you're sure you've got them right. Set out from one of the square or lozenge symbols; you have a choice of all four directions (except for those on the edge where we run out of room on the page). Each of the other lattice points, marked by a dot, already has a darker line passing through it, so when your pencil arrives at a dot you can only follow the one lighter lattic link as you must not go over the darker lines. When you've arrived at the next symbol, you can continue in any one of three directions, up, down or across, or lift your pencil and try following another route from one of the symbols, and so on until all the possibilities have been exhausted, all the lighter lines pencilled over and an astonishing Lattice Labyrinth has leaped from the page.

I don't want to spoil the joy of discovery, but as this procedure will probably be quite new to you I've made a start, shown in black. The lines of the completed lattice labyrinth form the boundaries of jigsaw-like pieces, the *supertiles*, which despite their intricate shape fit together perfectly to fill the figure, which could be extended indefinitely off the page. To enhance the appearance of *Japanese Lattice Labyrinth* (8,6) you can colour the supertiles in alternate shades (I've started one). The Japanese Labyrinths are not described until page 24, but think about the meaning of the symbols and why this is member (8,6) of the *family*.

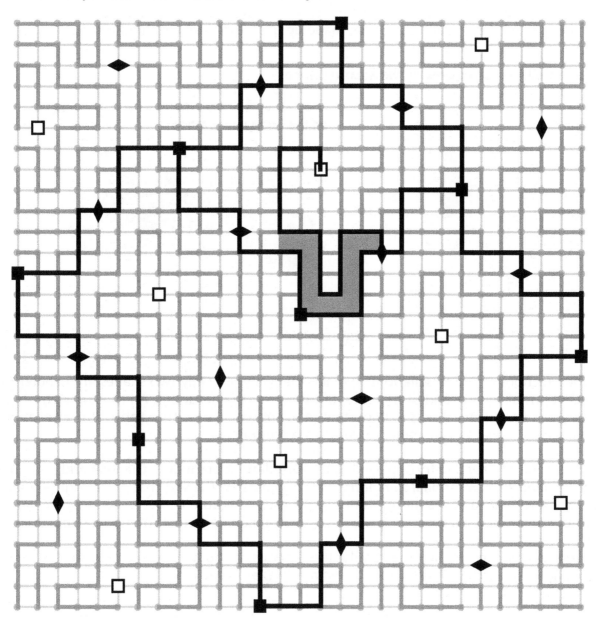

While I'll continue in the next few pages to explain some more of the basic concepts in WORDS, I recommend skipping backwards and forwards to try your HAND at another construction or two, starting on page 8. Mastering the techniques is what counts and will help you understand the theory that lies behind the practice.

Throughout the book, the first time (or two) that I use a word which might be unfamiliar, especially if I've made it up myself, I'll put it in **bold italics**. The meaning should often be clear from the context without much explanation.

You may know that there are only three **regular tessellations**, tessellations made up of just one size of **regular** (equal-sided and equal–angled) **polygons**. Of regular polygons, only squares, equilateral triangles and regular hexagons tessellate. All three regular tessellations can be laid out on the square or triangular lattice, as shown below. The chessboard tessellation of squares is very familiar but the tessellation of triangles is surprisingly rarely seen in decoration, games boards, tilings or pavings. Notice that although the triangles of the triangular tessellation are all of the same size and shape, half of them are upside down with respect to the other half. This means that, in order to make the whole tessellation we need to specify a repeat unit of two triangles, one in each orientation and together making a special **rhombus** with its shorter diagonal equal to its side. The tessellation of hexagons is familiar as the **Honeycomb** pattern. Notice that, in my jargon, each hexagon is actually a supertile made up of six individual triangular tiles.

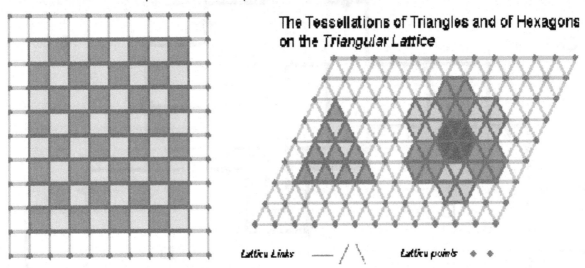

The Tessellation of Squares on the *Square Lattice*

The Tessellations of Triangles and of Hexagons on the *Triangular Lattice*

*Lattice Links* —— /  \    *Lattice points* ◆ ◆

I've shaded sections of the tessellations to make them clearer. Only two shades are needed for the tessellations of squares and of triangles, but three for the hexagons. All three patterns have a high degree of symmetry. This is so important in the discovery of Lattice Labyrinth Tessellations that we need to look at the symmetry of patterns more closely.

There are two forms of symmetry displayed by the tessellations of squares, triangles and hexagons – **mirror symmetry** and **rotational symmetry**. Mirror symmetry, the sort that makes right-handed people look left-handed in the mirror, is not our concern, because Lattice Labyrinths are hardly ever mirror symmetrical. They do all, however, display a high degree of rotational symmetry.

The **rotational symmetry** of the three regular tessellations is set out in the figure at the top of the opposite page, which shows the pattern of **symmetry axes** in each case.

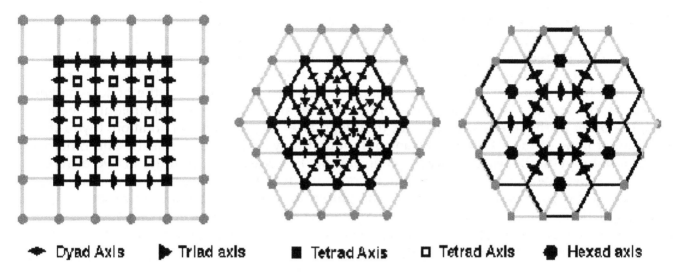

| ←•→ Dyad Axis | ▶ Triad axis | ■ Tetrad Axis | □ Tetrad Axis | ⬢ Hexad axis |

The axes of symmetry are represented by little lozenges and squares (as seen on the tessellation of page 5), triangles and hexagons. To see what these symbols represent it will help to look at simpler images – the stylised flowers illustrated below.

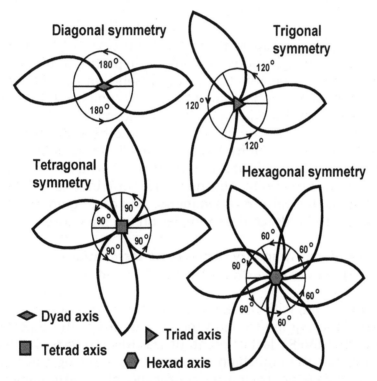

In the above figure you can see how each flower would appear the same if suitably rotated about an axis at its stem. The two-petalled flower has to be rotated through 180° before it appears unchanged, the petals falling exactly into the places vacated by each other. Two such rotations bring the petals back to their original positions, so we say that the flower displays two-fold, or **diagonal** rotational symmetry about a **dyad axis of symmetry**, represented by a lozenge. The three-petalled flower appears unchanged if rotated through 120° and three such rotations bring it back to its original position so we say it displays **trigonal** symmetry about a triad axis, represented by a triangle. The four-petalled flower appears unchanged if rotated through 90° and four such rotations bring it back to its original position so we say it displays tetragonal symmetry about a **tetrad** axis, represented by a square. The six-petalled flower appears unchanged if rotated through 60° and six such rotations bring it back to its original position so we say it displays **hexagonal symmetry** about a **hexad** axis, represented by a hexagon. The terms **dyad**, **triad**, **tetrad**, **hexad** have been coined from the Classical Greek for the numbers **2**, **3**, **4** and **6**. It's great to pay homage to Greek mathematicians such as Euclid and Pythagoras, who had already worked out much of the algebraic and geometric basis of this book more than two millennia ago.

If you've completed the tessellation on page 5, have another look at it. Imagine sticking a pin in the central small black square, for instance, and spinning the tessellation around it. You will see that the whole pattern displays tetragonal symmetry about that point. Imagining the tessellation to extend in all directions you will see that the same is true for every small black square and for every small hollow square. We have to distinguish between the solid- and the hollow-marked tetrad axes because they are situated differently – one set at the intersection of straight three-link lines, the other at the intersection of lines that bend after just one link. In addition, if you stick your pin into one of the lozenges you can confirm that they are each a dyad axis – this is harder to make out, but all the more satisfying when you do. I've drawn the lozenges in two different orientations because the tetragonal symmetry demands it – if you see what I mean (but don't worry about it if you don't). Now compare this pattern of axes with the pattern of squares and lozenges marking the symmetry axes of the tessellations of squares illustrated on page 7. Despite the increase in scale and the intricacy of the Japanese Lattice Labyrinth the way the symmetry axes are situated in relation to one another is just the same.

We can call this pattern of two sets of tetrad axes and one set of dyad axes the **442 pattern of rotational symmetry**. In this book, ALL the Lattice Labyrinth Tessellations based on the square lattice display this pattern of symmetry. That is a crucial and beautiful defining characteristic. You can confirm that it holds for the four examples on the back cover.

If you now take a look at the Lattice Labyrinth on the front cover of the book (or the related graph of green and red triangles and hexagons) and observe in particular the symbols representing hexad, triad and dyad axes, you may be able to assure yourself that they do indeed mark axes of hexagonal, trigonal and diagonal rotational symmetry. If you compare the pattern of these axes with that of both the tessellation of triangles and the Honeycomb tessellation shown on page 7, you will see that the patterns of axes are identical, though on different scales. In each case, every hexad is surrounded by six triads, each of which has three hexads for neighbours. Midway between each pair of hexads (and between each pair of triads) there lies a dyad axis. We can call this the **632 pattern of rotational symmetry**. In this book, ALL the Lattice Labyrinth Tessellations that are based on the triangular lattice display this pattern of symmetry. You can, again, check for yourself that this crucial defining characteristic holds for the examples on the back cover.

This **conservation** of the rotational symmetry of the **basic tessellations** of squares, triangles and hexagons is the first defining feature of Lattice Labyrinth Tessellations.

The second defining feature is that, with the exception of the centre-points of the hexagons and related supertiles, the tessellation graph forming the boundaries of the supertiles always passes through and connects together EVERY lattice point. (It was this property that led me to discover the Labyrinths.) This means that the branching corridors that make up the supertiles are never more than one square or triangle wide. Their arms or branches are as attenuated as possible, their boundaries are as long as possible, the supertiles are as deeply interpenetrating as possible. Lattice Labyrinth Tessellations have a rigorous purity and an extreme nature that give them both their theoretical and their visual appeal.

If Lattice Labyrinths were to be one day laid out as pedestrian pavings in public places they would constitute fertile playgrounds for the enjoyment and exploration of geometry and numbers by children young and old. I appeal to present and future architects among you – get them built. In the meantime, enjoy playing your way through this little book. Maybe you'll become obsessed, like me, and spend evenings searching for tessellations never before seen and new methods of finding them. More families of patterns are investigated and more techniques are described in my larger, more technical book on Lattice Labyrinth Tessellations (2014); I'm sure that striking new insights and abstract connections will arise in better trained minds than mine.

Lattice Labyrinths Tessellations based on the triangular lattice are the more beautiful and the more challenging, but we had better start with Labyrinths based on the square lattice, as these are much easier to construct and demonstrate the crucial techniques employed.

To reveal what can be done, here's just four supertiles of *Chinese Lattice Labyrinth* (15,12), turned through 45° to fit them on the page. I'll explain the numbers very soon.

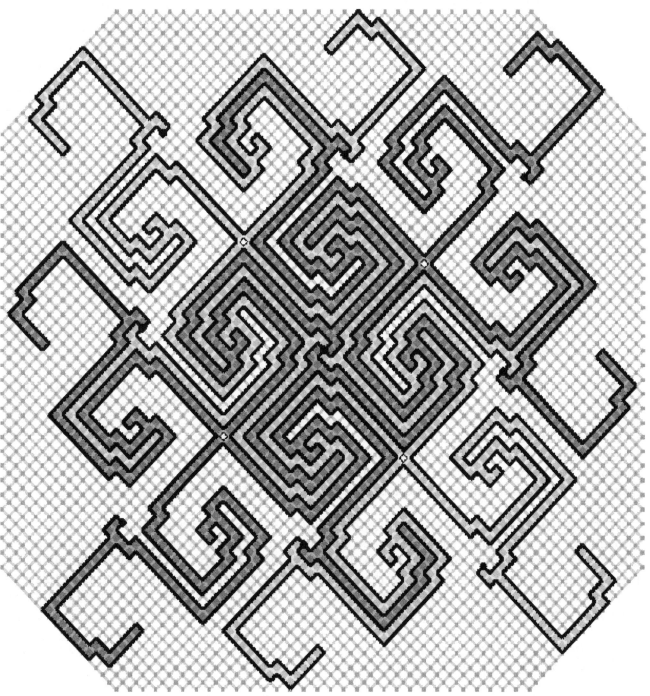

## Chinese Lattice Labyrinth (15,12)    Supertile Area $15^2 + 12^2 = 369$

The *Chinese Lattice Labyrinth* family of tessellations is the first I discovered, and the simplest to understand, so we'll start by constructing a couple of those – two different varieties of Chinese Labyrinth (7,6).

# Constructing Lattice Labyrinths From Scratch - Start Here!

To construct Chinese Lattice Labyrinth (7,6), we begin by setting out the array of symmetry axes on the square lattice. The pair of numbers (7,6) I call the **separation parameters** of the Labyrinth tessellation. They specify how far apart the axes lie. First of all, we'll do the standard construction shown on the top half of the page.

1. Starting near the left hand side of the page, we mark one of the lattice point dots with a small solid black square (representing a tetrad symmetry axis, see page 7). Count 7 links left to right across the lattice (in the *x* direction) and 6 links up (in the *y* direction) and mark another point with a solid square. Repeat this process until off the page. Now count 7 down and 6 across, or 7 up and 6 back-across from any of the marked squares and mark more lattice points. We've formed a big, tilted square **array of superlattice points**.

2. The second set of tetrad axes can now be marked using small hollow squares. They fall in the middle of each square of the array of superlattice points. These tetrad axes will turn out to lie at the centres of the supertiles of the Labyrinth tessellation.

3. Midway between the black squares or (it amounts to the same thing) midway between the hollow squares, mark the dyad symmetry axes. We find that each dyad axis falls at the centre-point of a lattice link.

4. Now for the crucial construction, the **COMPLEMENTARY GRAPH**, which we need to draw before we can draw the tessellation itself. Around each black-squared superlattice point we draw, in a light colour, a nest of three squares, of sides 2, 4 and 6, made up of lattice links. We stop at this because squares of side 8 would interfere with one another.

5. In between the nests around the superlattice points we have just room to draw nests of three squares, of side 1, 3 and 5, centred on each hollow-square marked tetrad axis.

6. Now we can see that every lattice point except the superlattice points lies on one of the squares of the complementary graph we have drawn, using up two of the four lattice links to that point. Only the superlattice points still have four unused lattice links intersecting at them. All complementary graphs of the Chinese Labyrinth family must have this property.

7. **NOW FOR THE MAGIC**. Using a contrasting colour, join up the superlattice points using only lattice links not covered by the lines of the complementary graph (hence the name), just as we did for the tessellation on page 5. The result is a **tessellation graph**, forming the boundaries of the supertiles of the Lattice Labyrinth tessellation. Shade the supertiles in two colours to bring out their shape, a "four-fold angular spiral". I've drawn and shaded just one – please complete the construction to see how they interlock.

8. I call the above construction, using nests of squares the **standard** construction, because it can readily be generalised to work for most Chinese Labyrinths and it's easy to draw.

9. The nests-of-squares construction is not the only possibility for Chinese Labyrinth (7,6) however. On the bottom half of the page I've set out an alternative, **non-standard** complementary graph including what I call **thumbed squares**. This generates a less obvious tessellation. I have begun drawing the tessellation graph and invite you to complete it by joining up the superlattice points using only those lattice links not covered by the lines of the complementary graph, thus revealing a labyrinthine tessellation.

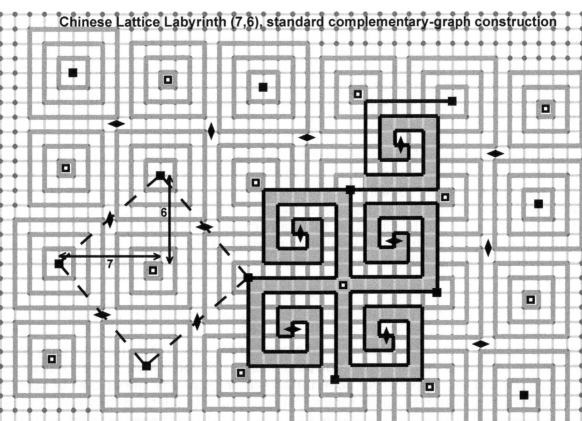

Chinese Lattice Labyrinth (7,6), standard complementary-graph construction

**Complementary Graph** ╌
**Tessellation Graph** ━ (all lattice links not used by the complementary graph)
**Superlattice Points (tetrad axes)** ■
**Supertile Centres (tetrad axes)** ▫    **Dyad Axes** ◆ ◈
  The dashed lines indicate one square of the array of superlattice points

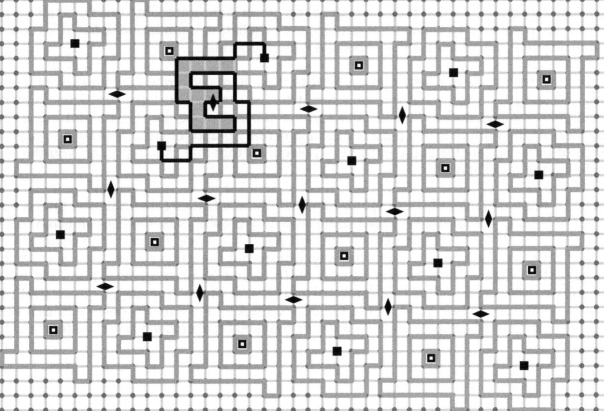

Chinese Lattice Labyrinth (7,6)  a non-standard construction employing thumbed squares

I've asserted that the nests-of-squares standard construction of a complementary graph employed for Chinese Labyrinth (7,6) can readily be generalised to work for MOST Chinese Labyrinths, that is for labyrinths with most values of a and b, the separation parameters. Remember that the separation parameters tell us how far apart to set out the superlattice points and hence all the symmetry axes. We can choose a and b as large as we like, so what do I mean by "most" of an INFINITE number of possibilities?

It turns out that a Chinese Labyrinth is theoretically drawable only for any pair of separation parameters of which one is **odd** and the other **even or zero**. As a convention, I always take **a** to be the odd parameter. See the "Small Print" on pages 22 and 23 for an explanation of why only these parameter pairs are possible. This is a *necessary* but not a *sufficient* condition for a Labyrinth tessellation to exist – we need to show that they can actually be drawn for all such cases. In fact, there is just one maverick case in which a Chinese Labyrinth cannot be constructed. We'll see later how that maverick redeems itself by proving to be the gateway into a whole new family, the *Serpentine Labyrinths*.

For a while, I thought that the standard complementary graph construction generalised from the (7,6) case would successfully generate a Labyrinth Tessellation for any pair of separation parameters. For any case there's only one way to construct the nests of squares so that they fit snugly together without overlap. **The rule is that for separation parameters (a,b) you draw a nest of (a-1)/2 squares, of side 2, 4, etc around each superlattice point tetrad axis and a nest of b/2 squares of side 1, 3 etc around the tetrad axis at each supertile centre.** You might like to select an (**odd,even**) parameter pair and try the rule out in pencil on page 45 of this book or on graph paper, or using software such as *Adobe Illustrator*©.

I've tried this rule out on case (**5,0**) on the page opposite. The standard complementary graph, consists of a nest of (**5-1**)/2 squares about each superlattice point and **0/2**, i.e. **none** about the superlattice centres. ALAS, when we draw in the graph of all the lattice links not used by this complementary graph, we find that the tessellation graph is separated into unconnected portions. Although we obtain a nice symmetrical pattern, the tessellation is not *monohedral*, it is comprised of three different shapes, tiles within supertiles within supertiles.

All is not lost for such (**a,0**) cases however. Next to the failure, I've drawn another complementary graph, including thumbed squares. I have left it for you to construct the resulting tessellation. You will find that we are saved. A proper Chinese Labyrinth emerges. How about higher-order (**a,0**) cases? At the bottom of the page is a complementary graph for (**11,0**), including squares with thumbs of differing lengths and widths fitting elegantly together. Again I've left you the fun of completing the tessellation.

Inspired by these examples of the (**a,0**) *sub-family* of Chinese Lattice Labyrinths, the time is ripe to look for a general rule for drawing their complementary graphs. With the (**5,0**) and (**11,0**) cases to guide you, have a try at (**7,0**) and (**9,0**) on the next page. I've made a start on the complementary graph of each, enough I hope to point the way. It turns out that the rule for cases (**5,0**), (**9,0**) and (**13,0**) is quite different from that for cases (**7,0**), (**11,0**) and (**15,0**). In general terms we need to distinguish between cases of the forms (**2c+1, 0**) and (**2c-1, 0**) where **c** equals **3, 4, 5**, etc. (Notice I've left out **c = 1** and **2**.)  To help you spot the trends and maybe formulate the general rules precisely, I've made a start on the complementary graphs for (**7,0**) and (**15,0**) on page 12 and for (**9,0**) and (**13,0**) on page 13.

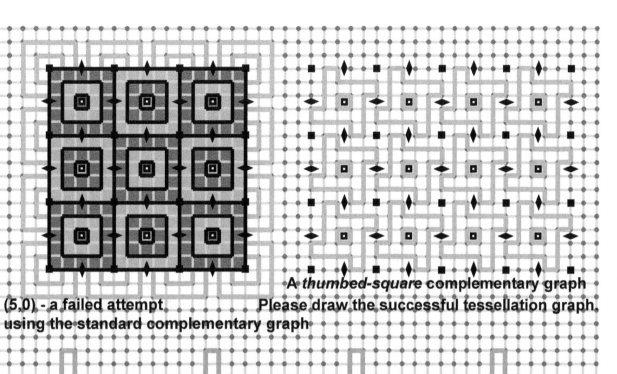

(5,0) - a failed attempt
using the standard complementary graph

A *thumbed-square* complementary graph
Please draw the successful tessellation graph.

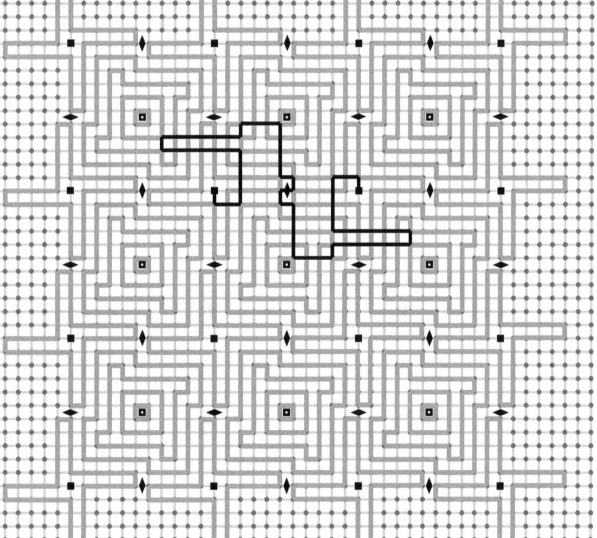

(11,0) - long-thumbed squares about the superlattice points
& nests of squares and short-thumbed squares about the supertyile centres
I've begun the construction of the labyrinthine tessellation graph.

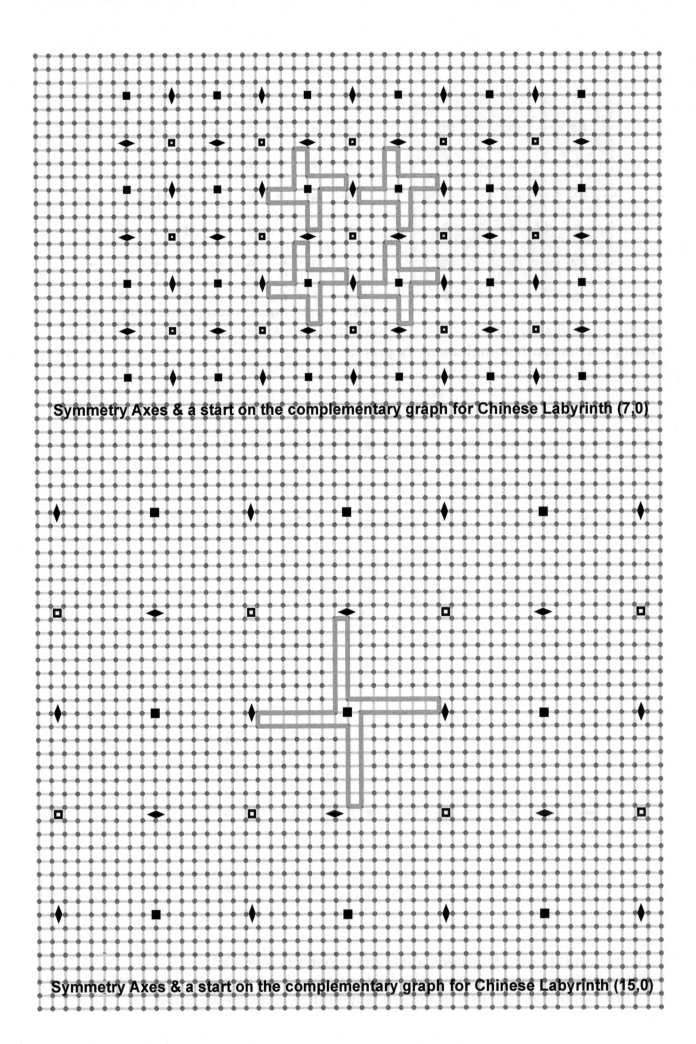

**Symmetry Axes & a start on the complementary graph for Chinese Labyrinth (7,0)**

**Symmetry Axes & a start on the complementary graph for Chinese Labyrinth (15,0)**

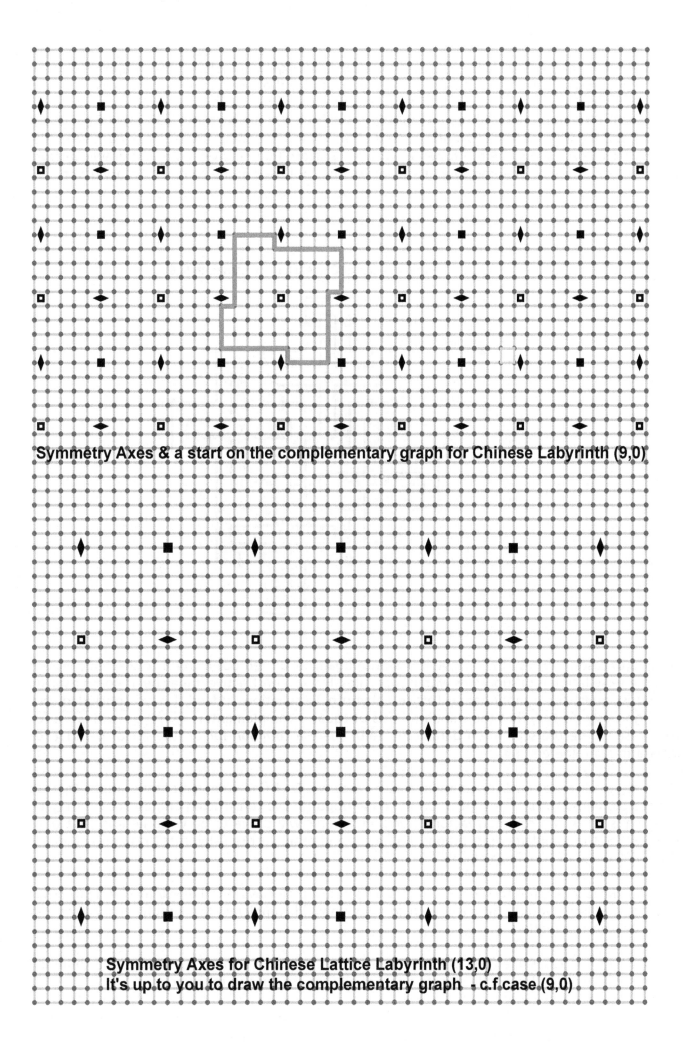

Symmetry Axes & a start on the complementary graph for Chinese Labyrinth (9,0)

Symmetry Axes for Chinese Lattice Labyrinth (13,0)
It's up to you to draw the complementary graph - c.f.case (9,0)

We have seen that there are fairly simple rules for constructing the complementary graphs of Chinese Lattice Labyrinths for the "usual" cases such as (7,6) and for the cases where **b=0**, such as (13,0) and (15,0), but there are still some awkward cases that fit none of these rules. For instance, suppose we try using the standard nests-of-squares rule for case (3,6). We are to construct nests of (3-1)/2 = 1 squares about each superlattice point and nests of 6/2 = 3 squares about the centre of each supertile. This construction is shown below, together with the pattern that emerges when we draw construct a tessellation graph by drawing over all the lattice links not used in the complementary graph.

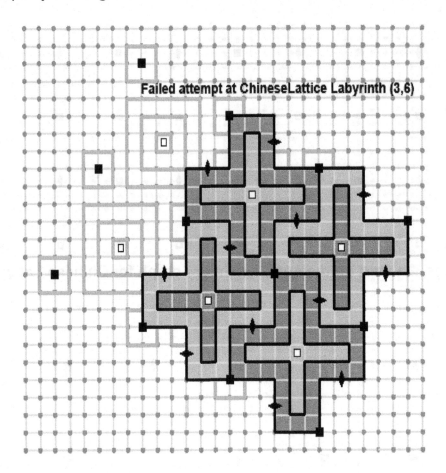

**Failed attempt at ChineseLattice Labyrinth (3,6)**

As in (a,0) cases, the tessellation graph has once again separated out into un-connected portions and the tessellation is of supertiles within supertiles. However attractive it may be, it doesn't satisfy my purist quest for a monohedral Lattice Labyrinth Tessellation. Indeed, I have concluded, but not rigorously proved, that the standard nests-of-squares complementary graph doesn't work for ANY case where the separation parameters (a,b) share a *common factor*, the factor being 3 for separation parameters (3,6).

Once more, all is not lost; for common factor cases we CAN find complementary graphs that do generate Chinese Lattice Labyrinths. On the page opposite I've constructed Chinese labyrinth (9,6), where 3 is again the common factor. Thumbed squares will prove to be versatile and fruitful as elements of the complementary graphs of Chinese and other Lattice Labyrinth Tessellation families that we'll come to later in this book.

Also on the page opposite I've set out the axes for you to have another go at case (3,6). With (9,6) for guidance, you should be able to complete a valid complementary graph for this case and construct the Tessellation.

Helpful hint – in cases like (3,6) where **b** is simply a multiple of **a**, nests of squares and thumbed squares drawn around **supertile centres only** are all that are needed for success.

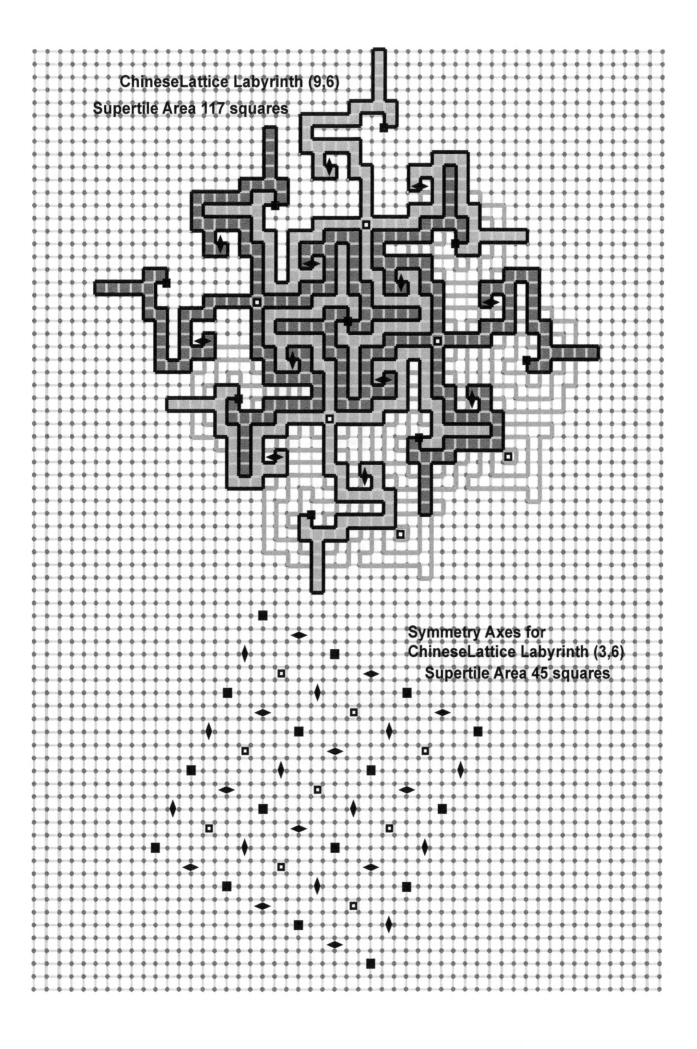

ChineseLattice Labyrinth (9,6)
Supertile Area 117 squares

Symmetry Axes for
ChineseLattice Labyrinth (3,6)
Supertile Area 45 squares

I trust that you've cracked the Chinese Labyrinth (**3,6**) on the previous page. A more spectacular case where **b** is a multiple of **a** is Labyrinth (**7,14**). Below, I've laid out the symmetry axes and have added just one thumbed square. You should be able to fill in "concentrically" the rest of this nest and, by repeating and fitting-in, draw the other nests of the complementary graph and so construct the pleasing Labyrinth tessellation.

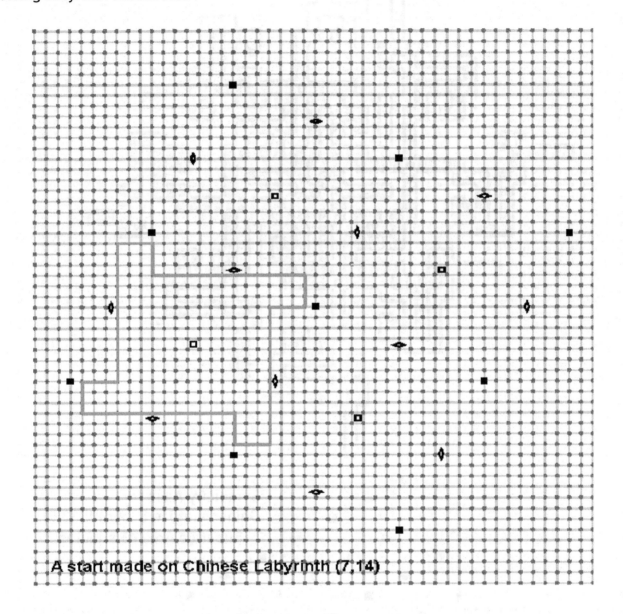

A start made on Chinese Labyrinth (7,14)

The crucial difference between the complementary graphs for cases (**3,6**) and (**7,14**) is the length of the thumbs. This indicates how we could construct (**5,10**), (**9,18**) and indeed any case of the sub-family (**a,2a**), all of which display a strong family resemblance one to another, despite the increasing number of elbows in the supertile arms, as **a** gets larger and larger and, indeed, tends to infinity. It turns out that the factor that determines the shape of the complementary graph is the HIGHEST common factor, **9** rather than **3** in case (**9,18**).

The complementary graphs for Chinese Labyrinths (**3,6**) and (**7,14**) consists of nests of squares and thumbed squares drawn around the supertile centres only. There are no nests centred on the superlattice points. Chinese Labyrinth (**9,6**), however, illustrated on page 13, has nests centred both on the supertile centres and on the superlattice points. We can make a good guess that nests are needed about both sets of tetrad axes when **a** and **b** share a common factor (always, of course an **odd** number) BUT **b** is not simply a multiple of **a**.

What has happened in case (**9,6**) is that the nests about the supertile centres remain the same as for case (**3,6**), but have been "forced apart" by the nests around the superlattice points, which snugly fill the spaces between the supertile-centre nests. With this in mind you should be able to complete the complementary graph and construct Chinese Lattice Labyrinth (**15,10**) on the lattice laid out below. I hope you don't mind my having made a start. Compare its shape with Chinese Lattice Labyrinth (**9,6**), illustrated on page 15.

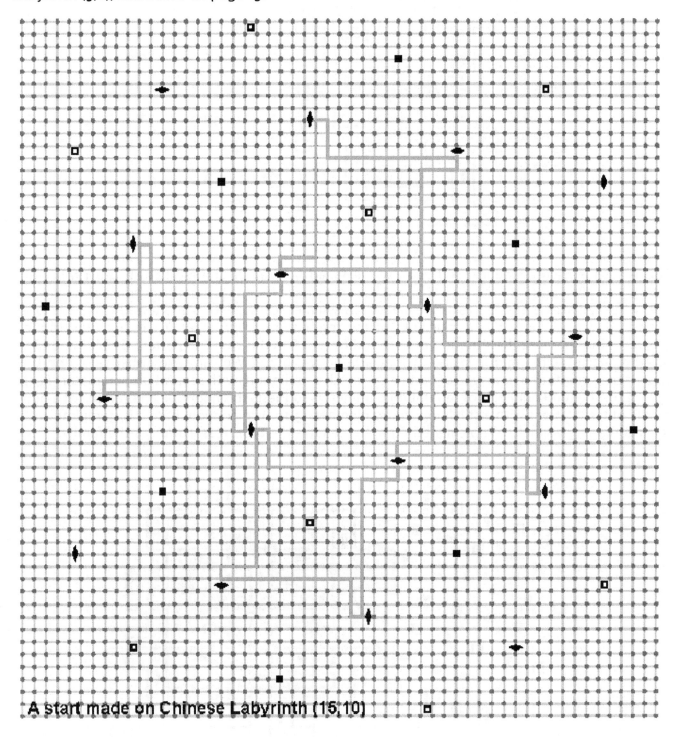

A start made on Chinese Labyrinth (15,10)

On page 8 I mentioned a maverick (**a,o**) case. You may already have discovered that the culprit is (**3,0**), for which it is impossible to construct a Chinese Lattice Labyrinth.

However, if we draw a complementary graph made up of three-by-one rectangles, we discover the pattern shown below, an elegant tessellation of capital letter I's.

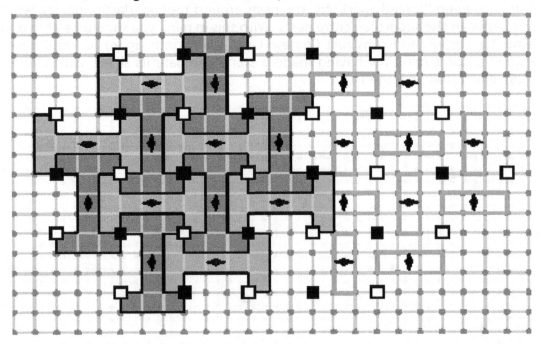

This is such a creative failure! Although we've failed to draw a Chinese Lattice Labyrinth with tetragonally symmetrical supertiles, the tessellation we have found still has the overall **442** symmetry but with dyad axes at the centre of the diagonally symmetrical supertiles and tetrad symmetry axes falling on lattice points where four links of the tessellation graph meet. There are two distinct sets of tetrads because the situation of each is different – in this case the pattern of lines and tiles around one set is the mirror image of that around the other. We can take either set to be the superlattice points of a new type of Lattice Labyrinth tessellation, with separation parameters **(3,3)**, NOT **(3,0)**. The **(odd,odd)** pair of separation parameters suggests this might be an example of a new family of tessellations with a still *lower-order* member, **(3,1)**. Sure enough, here it is, complete with an encouragingly simple complementary graph. I call this the *basic tessellation* of the family.

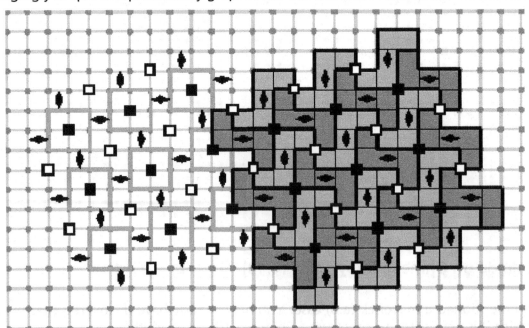

I call this family the *Serpentine Lattice Labyrinths*. Why "Serpentine"? If you complete the construction of the tessellations on the opposite page you will see the explanation.

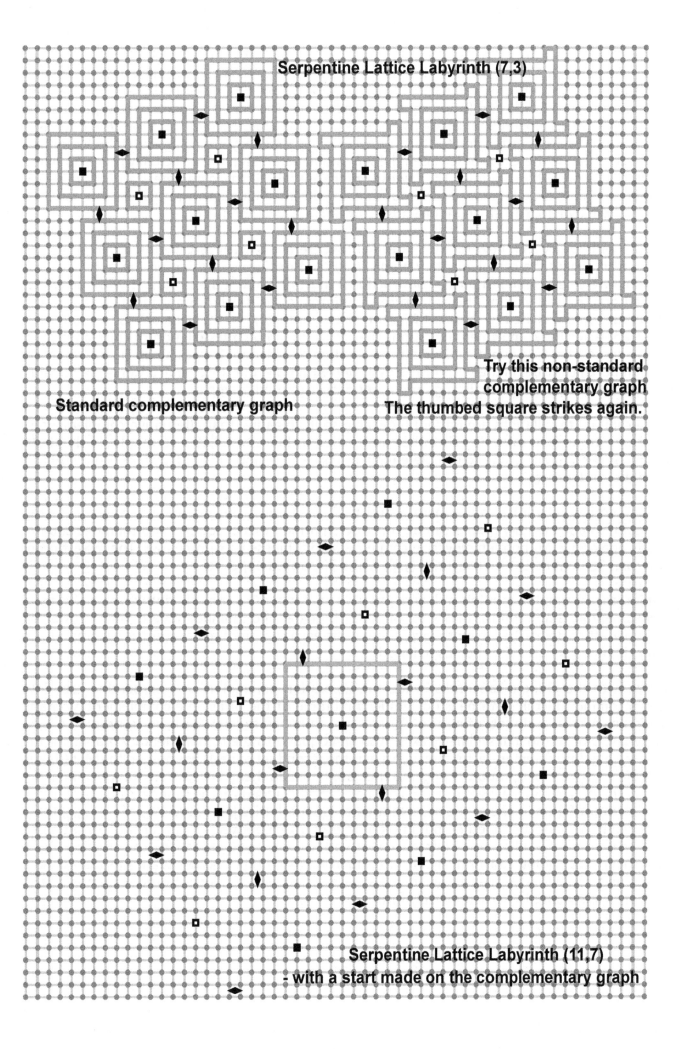

Serpentine Lattice Labyrinth (7,3)

Try this non-standard
complementary graph
The thumbed square strikes again.

Standard complementary graph

Serpentine Lattice Labyrinth (11,7)
- with a start made on the complementary graph

The Serpentine Labyrinth (3,1) illustrated on page 20 and the Labyrinths (7,3) and (11,7) completed (?) on page 21, can all be constructed using complementary graphs consisting of nests of squares centred on the two sets of tetrad axes. **In the standard rule for the *general case* (a,b) the two sets of nests consist of (a-1)/2 and (b-1)/2 squares respectively.** This rule is elegantly analogous to that for Chinese Lattice Labyrinth tessellations. There is a further analogy. In all three of the above cases, **a** and **b** don't share a factor greater than unity – we say that they are co-prime. (John Wykes has pointed out that in these low-order cases, the separation parameters are all actually prime numbers, so MUST obviously be co-prime! The *lowest-order* case where the parameters are co-prime but neither is prime is (25,9)) It indeed turns out that, as with the Chinese Labyrinths, the standard rule does not work when **a** and **b** are not co-prime but share a common factor.

To start with we need a new rule for cases (a,b) where a = b, such as Labyrinth (3,3), illustrated on page 20, where the complementary graph consists of rectangles centred on the dyad axes, the first time we've encountered such a construction. **I must go back and experiment with rectangles centred on the dyad axes of Chinese Lattice Labyrinths to see if valid complementary graphs can be found.** You are invited to construct Serpentine Labyrinth (7,7) on the page opposite, using nests of rectangles as the complementary graph. The result should be a tessellation of capital letter I's, akin to those of case (3,3) but curlicue-serifed. For higher-order a = b cases, (9,9), (11,11) and so on, the number of curlicues in the serifs increases by one each time, but the overall letter I-ness remains. I'm tempted to construct a really high-order, seriously curlicued example.

These capital I tessellations are rather predictable so it's pleasing that alternative complementary graphs for these (a,a) cases do exist. One construction includes castellated rectangles in the nests. Room has been left opposite to complete (5,5) in this way; analogous constructions can be utilised for higher-order cases.

For Serpentine Labyrinths (a,b) where **a** and **b** are not equal but do share a common factor, successful complementary graphs can be constructed using nests of squares and thumbed squares closely analogous to those employed in common-factor Chinese Labyrinth cases. The lowest order such case, Serpentine Lattice Labyrinth (9,3), is also illustrated opposite.

In this case, **a** is simply a multiple of **b** and, just as in analogous Chinese Labyrinth cases, a successful complementary graph consists of one set of nests of squares and thumbed squares only, centred on the superlattice points. In Labyrinth (9,3), these nests each contain three concentric squares and an outer thumbed square with thumbs of length 1.

Just as with the Chinese Labyrinths, when the separation parameters of a Serpentine Labyrinth have a common factor, but one is not simply a multiple of the other, the complementary graph consists of two separate sets of nests of squares and thumbed squares, each set centred on one of the two sets of tetragonal symmetry axes. Also analogously to the complementary graphs for the Chinese Labyrinth common-factor family, it is as if the second set of nests has pushed the first (simple multiple cases) set apart, snugly fitting into the gaps between them. To illustrate this phenomenon, I've drawn the symmetry axes of case (15,9) on page 24, and added a large thumbed square, which is the outermost figure in one of the nests centred on a superlattice point. Were this case (15,3) these nests would fit perfectly together, but in (15,9) there is room between them for broadly-thumbed nests centred on the other set of tetrad axes, as you will see if you repeat the construction about each superlattice point. Labyrinth (15,9), with supertiles made up of no less than **153** squares and a repeat unit of TWO supertiles, totalling **306** squares, is actually the lowest-order common-factor, but not simple-multiple, Serpentine Labyrinth!

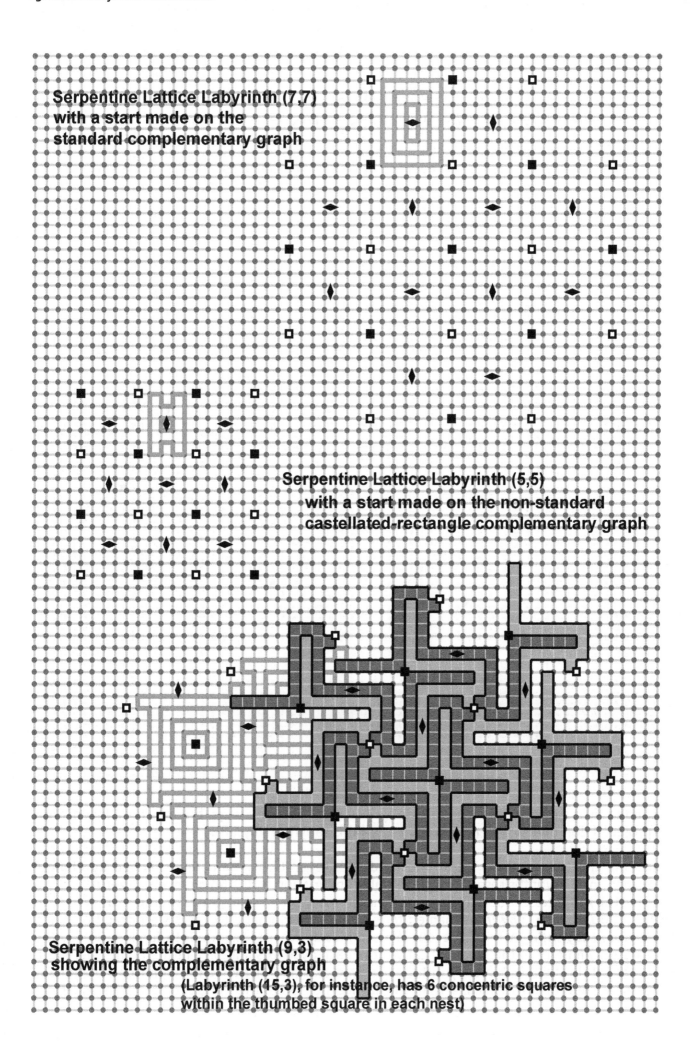

**Serpentine Lattice Labyrinth (7,7)
with a start made on the
standard complementary graph**

**Serpentine Lattice Labyrinth (5,5)
with a start made on the non-standard
castellated-rectangle complementary graph**

**Serpentine Lattice Labyrinth (9,3)
showing the complementary graph**
(Labyrinth (15,3), for instance, has 6 concentric squares
within the thumbed square in each nest)

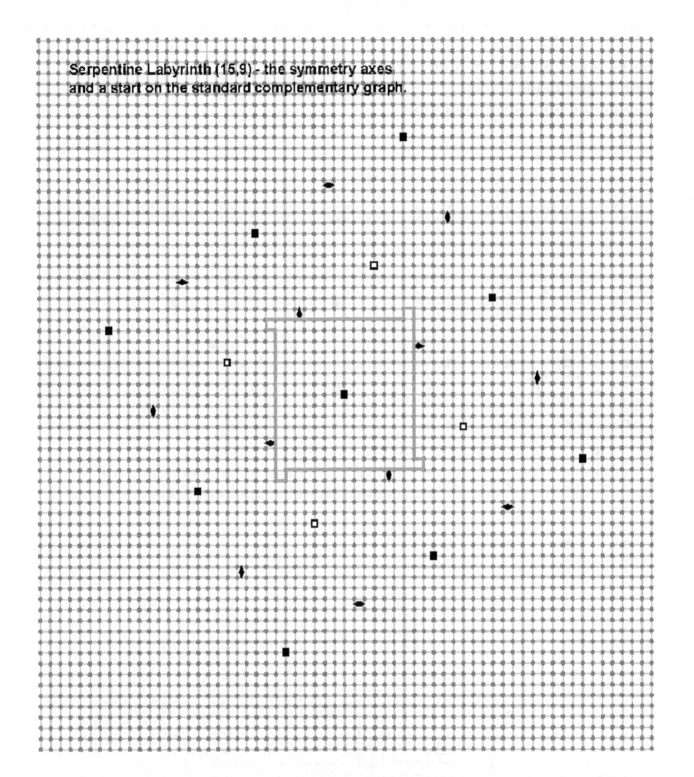

Serpentine Labyrinth (15,9) - the symmetry axes and a start on the standard complementary graph.

I've tried to fit 12 supertiles of Serpentine Labyrinth **(15,9)** onto the page, but may have failed, arms may project (I don't know; I've never constructed this Labyrinth myself).

Apart from the **(odd, even** or **zero)** and **(odd,odd)** separation parameters that specify Chinese and Serpentine Labyrinths respectively, the only combination left is **(even, even** or **zero)**. Can we construct Labyrinth tessellations for such parameter pairs? Page 25 illustrates examples of TWO such families, the *Japanese Labyrinths* (named to rebalance the homage to Far-Eastern traditions) and the *Scarthin Labyrinths* (named after the blessed hillside where I work). Both families display the familiar **442** symmetry. In the Japanese Labyrinths, each dyad axis falls on a lattice point where four links intersect. In the Scarthin Labyrinths, each dyad axis falls at the centre point of a diagonally symmetrical supertile. This means that the lattice points at the centre of Scarthin supertiles do not lie on the tessellation graph, a breach of the normal rule. I'm including this family in the Lattice Labyrinths because of the beautiful property that they and the Japanese Labyrinths neatly divide up all possible (even, even or zero) pairs of separation parameters between them. For any parameter pair you can draw either a Japanese or a Scarthin Labyrinth, never both. The complementary graphs of each family are made up of familiar elements and once again help us construct what can be bafflingly labyrinthine tessellations. Nevertheless, up to the time of writing, I have not solved the problem of discovering completely general rules for drawing complementary graphs for either family. Can you crack this problem?

I'm sorry to cut the treatment of Japanese and Scarthin Labyrinths so short, and to miss out Windmill and Sail Labyrinths altogether (see my longer book), but space is lacking. It's time for some optional mathematics summarising our tessellations on the square lattice.

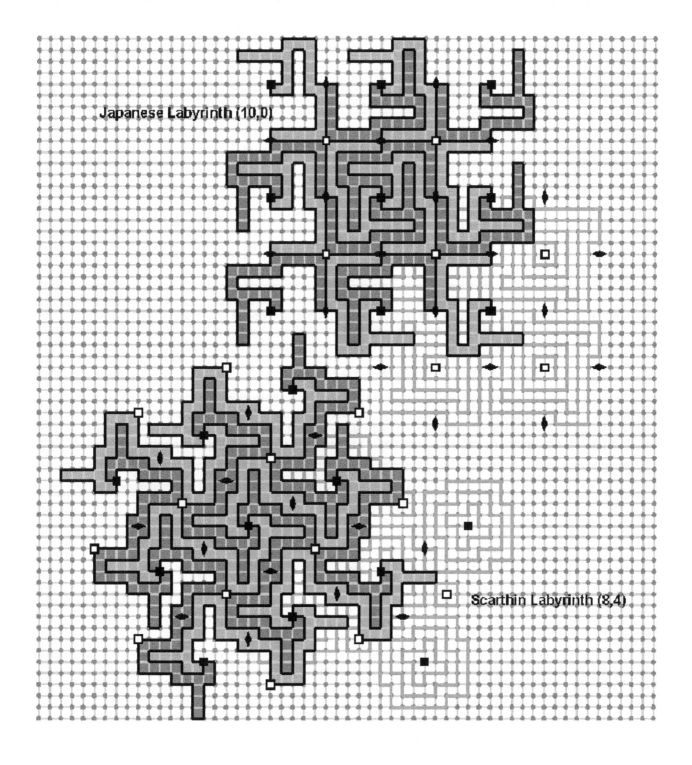

# The Small Print

Real mathematicians amongst you (I don't include myself in that family) will have noticed that the Lattice Labyrinth families mentioned are all tightly constrained by some not very advanced school-level algebra. There follows a skeleton summary.

The **CHINESE LABYRINTH** supertiles are made up of a central square and four identically-shaped arms and must be of the same area (**S**) as the giant squares that make up the array of superlattice points. This means that all cases must obey the following *Diophantine equation*, "Diophantine" meaning an equation where all the letters are substituted by *integers* only.

$$S = a^2 + b^2 = 1 + 4n$$

where **n** can be **zero** in the **basic**, or perhaps **degenerate** chessboard case, otherwise a **positive odd or even number**. Negative values of **a** and **b** would correspond to counting right-to-left, or up-to-down on the lattice, but just produce mirror images or rotations of the Labyrinths, so we can take both **a** and **b** to be positive without loss of generality. You can check that the Diophantine equation simply implies what we took for granted when constructing Chinese Labyrinths, that of **a** and **b** one must be **odd**, the other **even or zero**. All such pairs give solutions of the equation. See the table of low-order cases below.

| Superlattice Point Spacing | | n | S |
|---|---|---|---|
| a | b | Area of each arm | Chinese supertile area |
| 1 | 0 | 0 | 1 (chessboard) |
| 1 | 2 | 1 | 5 |
| 3 | 0 | (2) not possible | (9) can't draw |
| 3 | 2 | 3 | 13 |
| 1 | 4 | 4 | 17 |
| 3 | 4 | 6 | 25 |
| 5 | 0 | 6 | 25 |
| 5 | 2 | 7 | 29 |
| 1 | 6 | 9 | 37 |
| 5 | 4 | 10 | 41 |
| 3 | 6 | 11 | 45 |
| 7 | 0 | 12 | 49 |
| 7 | 2 | 13 | 53 |

**SERPENTINE LABYRINTH** supertiles are diagonally symmetrical about a central square, so each supertile must contain an odd number of squares. To each square of the array of superlatttice points correspond TWO supertiles at right angles to each other, so that the area of the repeat unit (**R**) that needs to be repeated to make the potentially infinite tessellation is **twice** the area of a supertile (**S**) and the Diophantine equation to which all cases must conform is:

$$R = 2S = a^2 + b^2 = 2n$$

where **n** is a **positive odd number**. It is not hard to prove that **a** and **b** must both be odd, and, once again, all such pairs are possible. A table of the lowest-order cases is given below. A beautiful relationship between the Chinese and the Serpentine tables is evident.( In theory (**1,1**) might be included as a degenerate Serpentine Labyrinth, one white plus one black square of the chessboard being the repeat unit of area **2**, but I prefer to consider (**3,1**) as the lowest-order **basic** case of this family.)

| Superlattice Point Spacing | | R | S |
|---|---|---|---|
| a | b | Serpentine repeat unit area | Serpentine supertile area |
| 3 | 1 | 10 | 5 |
| 3 | 3 | 18 | 9 |
| 5 | 1 | 26 | 13 |
| 5 | 3 | 34 | 17 |
| 5 | 5 | 50 | 25 |
| 7 | 1 | 50 | 25 |
| 7 | 3 | 58 | 29 |
| 7 | 5 | 74 | 37 |
| 9 | 1 | 82 | 41 |
| 9 | 3 | 90 | 45 |
| 7 | 7 | 98 | 49 |
| 9 | 5 | 106 | 53 |

A satisfying bonus is to have discovered numerical relationships like $9^2 + 1^2 = 2 (5^2 + 4^2)$ by means of geometry. **JAPANESE LATTICE LABYRINTHS** are characterised by a repeat unit made up of identically shaped supertiles in FOUR different orientations successively at right-angles to each other and each made up of an odd number of square tiles (why an ODD number? I leave that puzzle for you to solve). The Diophantine equation that must be obeyed is:

$$R = 4S = a^2 + b^2 = 4n \quad \text{where } n \text{ is an } \textbf{odd} \text{ number}$$

A table of the lowest-order cases of the Japanese Lattice Labyrinths is given at the top of the opposite page.

| Superlattice Point Spacing | | R | S |
| a | b | Japanese repeat unit area arm | Japanese supertile area |
|---|---|---|---|
| 4 | 2 | 20 | 5 |
| 6 | 0 | 36 | 9 |
| 6 | 4 | 52 | 13 |
| 8 | 2 | 68 | 17 |
| 8 | 6 | 100 | 25 |
| 10 | 0 | 100 | 25 |
| 10 | 4 | 116 | 29 |
| 12 | 2 | 148 | 37 |
| 10 | 8 | 164 | 41 |
| 12 | 6 | 180 | 45 |
| 14 | 0 | 196 | 49 |

The identical series of supertile areas in all the three (Chinese, Serpentine, Japanese) Lattice Labyrinth tables above arises because from the superlattice points of Chinese Labyrinth (1,2), with repeat unit and supertile area 5, we can select the superlattice points of Serpentine Labyrinth (3,1) with repeat unit 10, but supertile area still 5 and then, furthermore, from the superlattice points of the Serpentine Labyrinth (3,1) we can select the superlattice points of Japanese Labyrinth (2,4) with repeat unit 20 but with supertile area still equal to 5. I've tried to show how this works in the figure below.

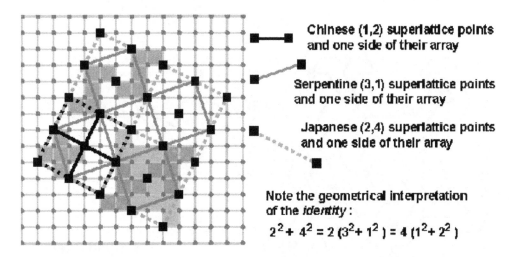

Chinese (1,2) superlattice points and one side of their array

Serpentine (3,1) superlattice points and one side of their array

Japanese (2,4) superlattice points and one side of their array

Note the geometrical interpretation of the *identity*:

$$2^2 + 4^2 = 2(3^2 + 1^2) = 4(1^2 + 2^2)$$

The **SCARTHIN LATTICE LABYRINTHS** are characterised by a repeat unit made up of identical diagonally symmetrical supertiles in TWO different orientations at right-angles to each other. The number of tiles in each supertile, S, is a multiple of four. (Once more, I leave you a puzzle to solve. Why is this so?) The Diophantine equation that must be obeyed is:

$$R = 2S = a^2 + b^2 = 4n \quad \text{where n is an \textbf{even} number}$$

The solutions to this equation are just those (**even, even or zero**) separation parameter pairs absent from the table of possible Japanese Labyrinths. With a bit of head scratching and time you may be able to work out why Japanese Labyrinths cannot be drawn for Scarthin Labyrinth separation parameters and vice versa.

| Superlattice Point Spacing | | Scarthin repeat unit area | Scarthin |
| a | b | R | Supertile area S |
|---|---|---|---|
| 4 | 0 | 16 | 8 |
| 4 | 4 | 32 | 16 |
| 6 | 2 | 40 | 20 |
| 8 | 0 | 64 | 32 |
| 6 | 6 | 72 | 36 |
| 8 | 4 | 80 | 40 |
| 10 | 2 | 104 | 52 |
| 8 | 8 | 128 | 64 |
| 10 | 6 | 136 | 68 |
| 12 | 0 | 144 | 72 |

**END OF SMALL PRINT**, which you don't need to have grasped in order to draw the pictures! I haven't room to mention the *Windmill Labyrinths*, a family drawable for ALL (**even, even or zero**) pairs of separation parameters, or their dissection into *Sail Labyrinths*. The complementary graphs of Windmill Labyrinths are hard to find as they are also Windmill Labyrinths!

# Beautiful Labyrinth Tessellations
# on the Triangular Lattice

Back to the BIG PRINT and on to the even greater wonders of Labyrinth tessellations on the triangular lattice. Do they exist? They certainly do, and here is an example of a *Trefoil Lattice Labyrinth*. I've commissioned some T-shirts printed with this lively tessellation, but so far I've had to give them away — I think the T-shirt was an unfashionable shape.

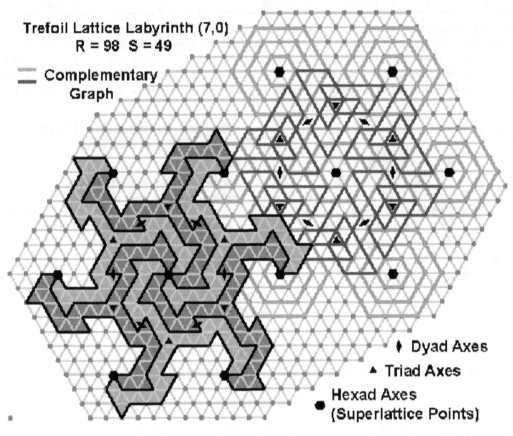

Trefoil Lattice Labyrinth (7,0)
R = 98  S = 49

Complementary Graph

Dyad Axes
Triad Axes
Hexad Axes
(Superlattice Points)

This tessellation exhibits a high degree of rotational symmetry. Assuming it's extended for ever and ignoring the shading, you can see that each hexagon symbol marks a *hexad* axis of symmetry, about which a rotation of 60° leaves the pattern looking unchanged and six such rotations make a complete rotation. At the centre of each supertile of the tessellation, a triangle marks a *triad* axis of symmetry. Not just the supertile but the whole tessellation appears unchanged if rotated through 120°; three such rotations bring you back to the start. Midway between adjacent hexad axes and triad axes, *dyad* axes are to be found at the centre point of lattice links, the centre points of each *side* of the supertiles. We can say that the whole tessellation exhibits **632** rotational symmetry, the same pattern of rotational symmetry as displayed by the triangular lattice itself, shown below.

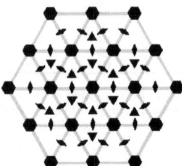

Is there an infinite family of Trefoil Labyrinths; if so, how do we discover them?

We can learn a lot from the practical task of constructing another Trefoil Lattice Labyrinth.

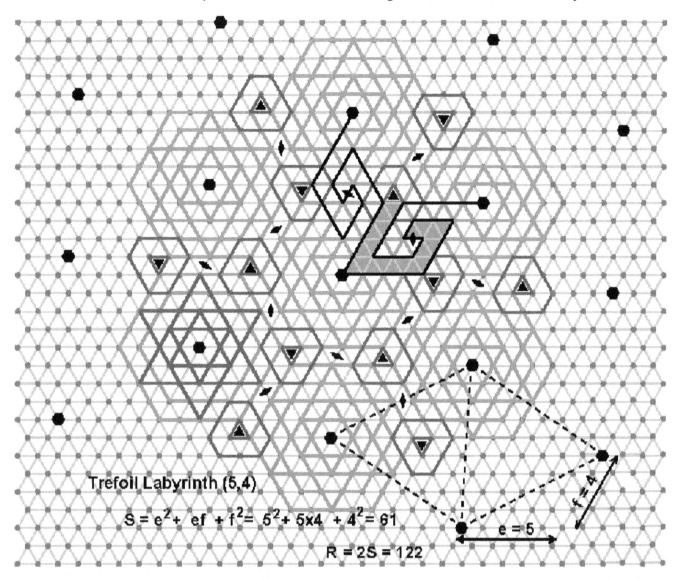

Trefoil Labyrinth (5,4)

$$S = e^2 + ef + f^2 = 5^2 + 5 \times 4 + 4^2 = 61$$

$$R = 2S = 122$$

The superlattice points are set out at separation **(5,4)** as shown above, forming a large tilted triangular array. The area of a supertile equals that of a triangle of the array. Can you show that the area of each triangle, measured in triangles of the original lattice is given by:

$$S = 5^2 + 5 \times 4 + 4^2 = 61$$

For Trefoil Labyrinths, I'm afraid we can't confine Diophantine Equations to the small print. We need to know which separation parameter pairs **(e,f)** are possible. Each supertile has a central triangle and three arms of identical shape, so the equation that they must obey is:

$$S = e^2 + ef + f^2 = 1 + 3n$$

Where **n** is a positive integer (or **zero** in the basic case, the tessellation of triangles). To generate the whole tessellation, we need to repeat an area corresponding to the TWO triangles shown by dotted lines, so the repeat unit of the tessellation **R** is equal to twice S.

I have drawn some of the complementary graph for case **(5,4)**. It consists of *hexagrams* **of the first and second order** (stars of David) and hexagons of side **3** and **4** nested about each superlattice point, and a triangle of side **1** and a *blunted triangle* nested about each supertile centre. The complementary graph needs to use up four of the six lattice links meeting at each lattice point other than the superlattice points. I've started the tessellation graph. The eye can't help interpreting (**e,e-1**) Labyrinths such as Trefoil (**5,4**) as three–dimensional banks of steps, though close inspection reveals an Escher-like paradox. Nests of hexagrams and hexagons around the superlattice points and of triangles and blunted triangles around the supertile centres make  successful complementary graphs for all (**e,e-1**) Trefoil Labyrinths. It's worth checking that all (**e,e-1**) cases satisfy the Diophantine equation. In fact, it's easy to prove that, taking **e** to be greater than or equal to **f** for convenience, **the only cases  which DON'T satisfy the Diophantine equation are those where e = f , f + 3, f + 6. etc. In general, Trefoil Labyrinths CANNOT be drawn for separation parameters (e,f) where e = f + 3r,  r being zero or a positive integer.**

We haven't proved that Trefoil Labyrinths can actually be drawn for all other separation parameter pairs (**e,f**) – we have to find out how to construct them first. Let's look at some examples. The complementary graphs for all allowed cases of the sub-family (**e,1**) turn out to be both simple and beautiful. To construct the complementary graph we simply draw a nest of hexagons of increasing size around each superlattice point until we run out of room. We then draw a nest

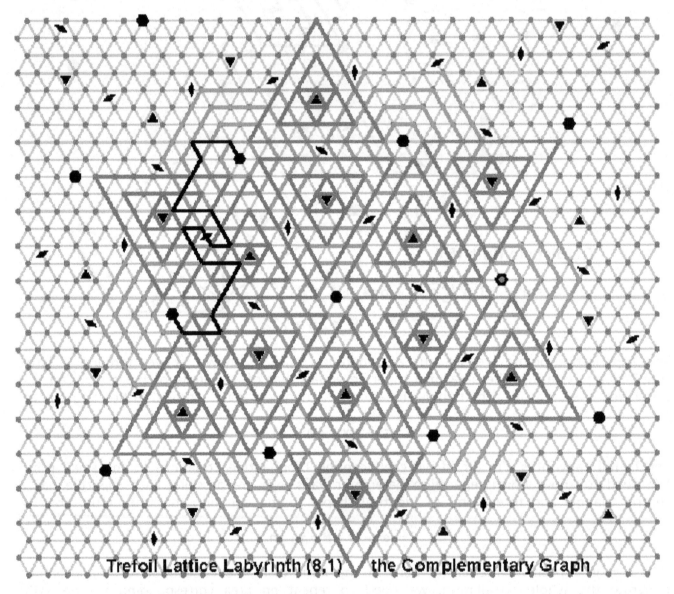

Trefoil Lattice Labyrinth (8,1)    the Complementary Graph

of triangles and inverted triangles about each supertile centre to fill the spaces between the hexagons and finally add more triangles until we run out of room for more and the corners of the largest triangles just reach the innermost hexagons. This method will work for any any (**e,1**). I hope you don't mind me starting the tessellation graph. Trefoil Labyrinth (8,1) can be seen as not only a member of sub-family (**e,1**) but also of sub-family (**e, e-7**). Does this give us any help in the search for the complementary graph for cases such as (**10,3**)? Maybe it narrows the search for the beautiful solution below. I've set out the axes and a complementary graph, with just a hint of the tessellation graph that outlines six-armed (but trigonally not hexagonally symmetrical) supertiles.

**Complementary Graph**

**Trefoil Lattice Labyrinth (9,5)**　　　　　　　　R = 152　S = 76

Nerd that I am, I can spend contented hours absorbed in the search for the complementary graphs of Trefoil Lattice Labyrinths previously "seen only by God". Below are some of the elements that we need to put together to construct successful complementary graphs.

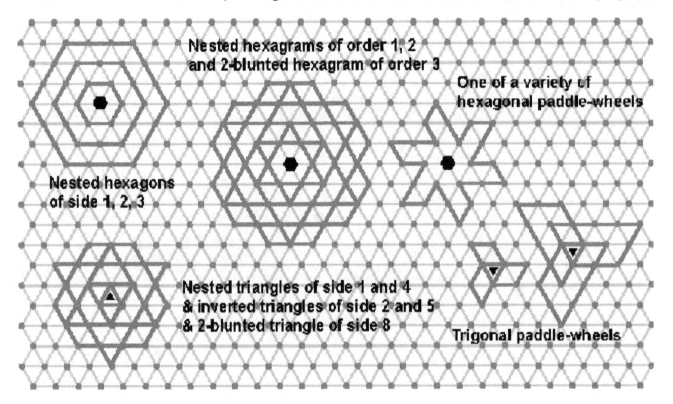

The elements of a complementary graph must be hexagonally symmetric about the superlattice points and trigonally symmetric about the supertile centres; the symmetry about the dyad axes then emerges automatically. The simplest hexagonally symmetrical figure is the hexagon, and nests of these often contribute to a complementary graph, but as they use up only two links to each lattice point, trigonally symmetrical elements will need to intersect all the hexagons. Nested hexagrams intersect each other, at leaving only points on the boundary of the nests to be reached by the trigonal figures. The trigonally symmetrical elements may consist of nests of equilateral triangles in the same orientation, but then need to be deeply penetrated by the hexagonally symmetrical figures. If nested with inverted triangles, as shown above, only some points on the boundary will need to be reached by the hexagonal figures. The final feature to point out is that sometimes the hexagrams and triangles need to be **blunted**, by one or more lattice link, as shown. Forearmed with experience of other related cases we can usually arrive at complementary graphs fairly quickly by juggling nests of hexagons, hexagrams, triangles and inverted triangles, blunted where necessary. I can enjoy, or endure, hours of such juggling – if crowned with success. Finally, what about those **paddle wheel** elements shown above?

Paddle wheel elements I don't find aesthetically satisfying but for (**e,o**) cases we do seem to need them, for instance Trefoil Labyrinth (**8,o**), which you are invited to complete on the opposite page. Finding a valid complementary graph, involving **hexagonal paddle-wheels**, took me most of a day. The paddle-wheel elements are analogous to the eared squares we needed for square lattice families. The common feature shared by all Lattice Labyrinth cases for which thumbed-square or paddle-wheel elements are required is that in these cases the complementary graph seems to require some extra "spiral" character to remove or reduce the contribution of **mirror symmetrical** elements such as nests of hexagons or hexagrams, one half of which "simply reflect" the other. More study needed.

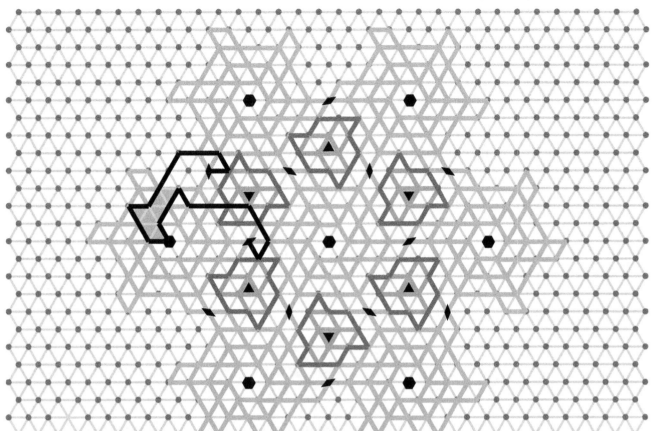

**Trefoil Labyrinth (8,0)**
**- found after a long struggle with various paddle-wheel constructions**

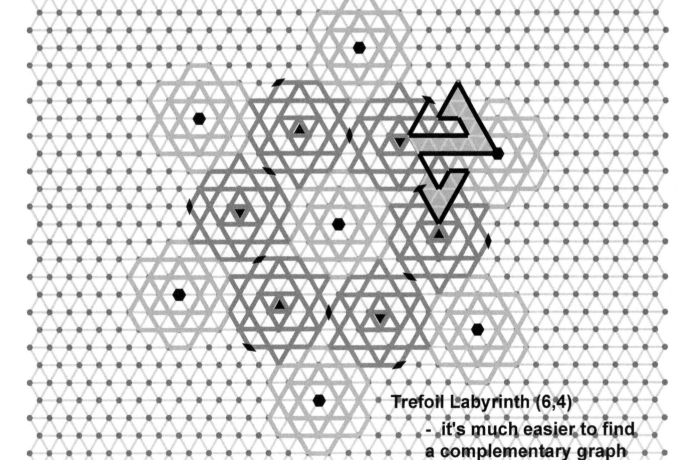

**Trefoil Labyrinth (6,4)**
**- it's much easier to find**
**a complementary graph**

I am loath to leave the Trefoil Labyrinths, which I consider to be the Queen of the Lattice Labyrinths families. The intersecting nests of figures which make up their complementary graphs are beautiful in themselves, and finding them is a challenge requiring insight and ingenuity. Whether completely general rules for drawing them can be compiled I do not know. I have worked out rules for sub-families of the form (**e,c**) and (**e,e-c**), where **c** is kept constant, for all values of **e** and SOME values of **c**, but the number of such families is infinite, so generalising from sub-families to all cases may be an intractable problem.

Reluctantly, I'll finish the Trefoil Labyrinth section by revisiting an exercise I set myself, to construct an example new to me of as high an order as can be fitted onto a page of the book. I chose case (**9,5**) and it took me about two hours to crack the complementary graph problem. First of all I tried drawing nests of simple hexagons – there's room for a nest of six around each superlattice point. In between, around each triad axis (supertile centre) I then constructed a nest of triangles. There has always to be at least an inner triangle of side **1**, in this case I found room for just one inverted triangle of side **2**, then further triangles of side **4** and side **7** and a triangle of side **10**, *2-blunted*. These didn't reach far enough to intersect with the hexagons of side **1** and side **2**, so I replaced the latter with hexagrams of order **1** and **2**, which fitted nicely. Things were looking promising, but a problem remained.

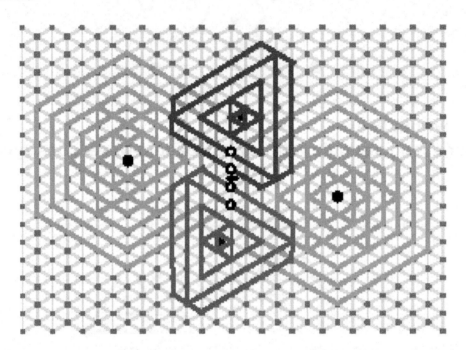

Four lattice points, ringed in the diagram, are reached once only; there's no way of reaching them a second time without using up ALL the lattice links to other lattice points. **Failure.** My next strategy was to replace the hexagons of side **1, 2, 3 and 4** by hexagrams of order **1, 2, 3 and 4**, which just fitted together, but I still could not construct figures about the triad axes that would intersect the hexagonally symmetric figures at all the necessary points. **Second failure.** Finally (as it turned out), I decided to start at the triad axes, around which I made nests of triangles as large as there was room for, finishing with triangles of side **7** and inverted triangles of side **8** (*2-blunted*) and **11** (*3-blunted*). This left room around each superlattice point for hexagrams of order **1** and **2** and a *1-blunted* hexagram of order **3**. **Success!** As elegant a solution as I had hoped for, but by no means obvious. I sometimes find that even such lovely complementary graphs do not work, the tessellation graph generating a pattern of supertiles within supertiles, akin to the square-lattice example on page 11. This graph, however, didn't let me down. I leave you the satisfaction of drawing this new Trefoil Lattice Labyrinth – a tessellation of character.

Complementary Graph

Trefoil Lattice Labyrinth (9,5)

R = 152  S = 76

The only one of the three regular tessellations that occurs naturally on a scale visible to the naked eye is that of hexagons, evolved as the honeycombs of bees and approximately generated by the cooling of volcanic basalt sheets, as at the Giant's Causeway on the Antrim coast of Northern Ireland.

Can we find a family of Labyrinth Tessellations based on the honeycomb? Looking for these, I was held up by the false assumption that such tessellations must themselves be made up of hexagons. In fact, the **632** rotational symmetry of the honeycomb is shared by Lattice Labyrinth Tessellations the hexagonally symmetrical supertiles of which are made up of triangles, as are hexagons themselves.

The figures on the opposite page show the honeycomb itself and three examples from the ***Honeycomb Lattice Labyrinth*** family. For the first time we need three different shades to distinguish the supertiles so that no two of the same colour meet along an edge. All four tessellations do indeed share the **632** pattern of symmetry axes. The hexad axis superlattice points are situated at the centre of each supertile, the triad axes where three supertile boundaries meet, the dyad axes at the centre of each ***side*** (see fig. opp.) of each supertile.

The area of a supertile is equal to the area of a ***rhombus*** as shown in the figures by dotted lines, and this in turn contains TWO equilateral triangles each of area $e^2 + ef + f^2$ where **(e,f)** are the separation parameters, measured out just as for the Trefoil Labyrinths. So in this Family, the repeat unit of the tessellation, **R**, is equal to the supertile area **S**, which must be divisible by **6** to maintain the hexagonal symmetry of the supertile, so that

$$R = S = 2(e^2 + ef + f^2) = 6n$$

and therefore

$$e^2 + ef + f^2 = 3n \text{ where } \mathbf{n} \text{ is a positive integer.}$$

This means that superlattice point separations possible for Honeycomb Labyrinths have **e = f + 3r** where r is zero or a positive integer, which comprise all the parameter pairs not possible for Trefoil Labyrinths.

As Honeycomb Labyrinth superlattice points do not lie on the tessellation graph, their complementary graphs (except for the Honeycomb itself) must include a cartwheel figure centred on each superlattice point. Three links must meet symmetrically at each triad axis. The dyad axes can each lie either at a lattice point or at the mid point of a lattice link. When the dyad axes lie on lattice points, the complementary graph must symmetrically use up four links to each, but when the dyad axis lies at the centre point of a lattice link the complementary graph must symmetrically avoid that link. You might like to think about the values of **(e,f)** which correspond to these two alternative situations. The complementary graphs of Honeycomb Lattice Labyrinths can be difficult or impossible to resolve into the shapes that are familiar from the Trefoil Labyrinth family. I've used different shades of line in the figures opposite to distinguish some of their features.

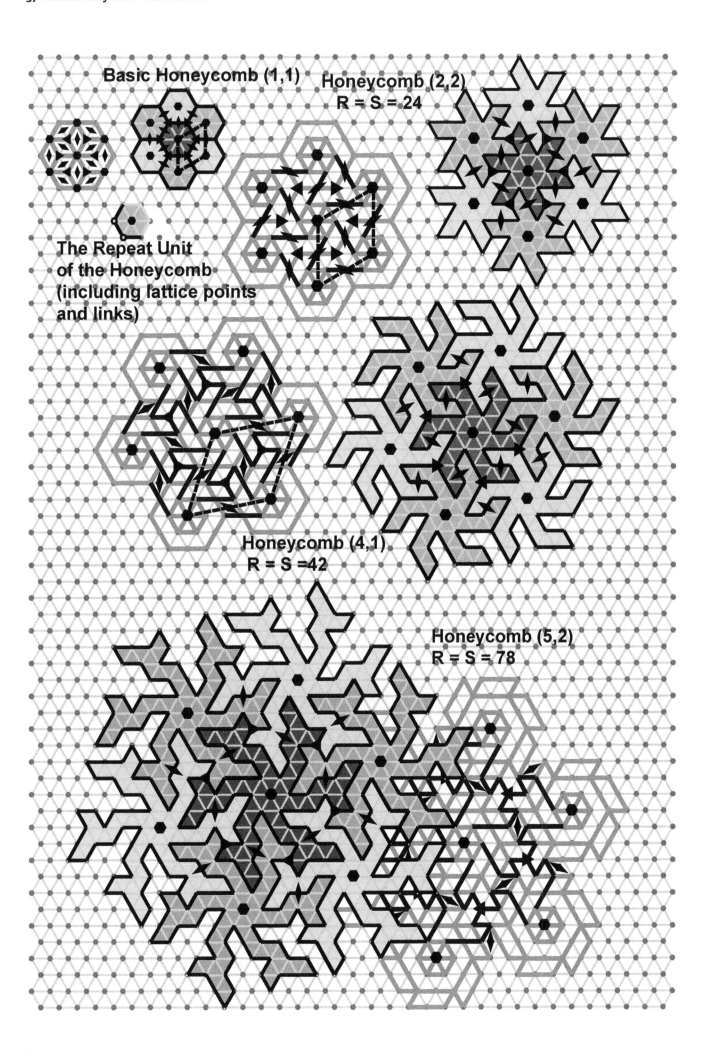

Basic Honeycomb (1,1)

Honeycomb (2,2)
R = S = 24

The Repeat Unit
of the Honeycomb
(including lattice points
and links)

Honeycomb (4,1)
R = S = 42

Honeycomb (5,2)
R = S = 78

On the next two pages, I've illustrated complementary graphs for two somewhat higher-order Honeycomb Lattice Labyrinths. I've left the tessellation graphs entirely to you.

The first case is Honeycomb Labyrinth (9,3). The complementary graph is exquisitely satisfying. After we've drawn in the obligatory cartwheels, we add hexagons of side 2, 3 and 4 to each nest. In between these nests there is just room for nests of triangles of side 3 and 6 centred on the trigonal axes. Finally, centred on each trigonal axis we add what I call a trigonal spiral figure, which, after three elbows just reaches the three nearest cartwheels. Fortunately, this elegant complementary graph does indeed generate a valid Labyrinth tessellation. There should just be room on the page to complete a central supertile and its six surrounding neighbours. Their spikeyness endows them with an assertive character.

This complementary graph of Honeycomb Labyrinth (9,3), with its elegant trigonal spirals, can be generalised for (e,3) cases, but though it works for (6,3), it does not work either for (3,3) or for (12,3), but produces "tiles within tiles". How about Honeycomb (15,3)?

Honeycomb Lattice Labyrinth (9,3)
- complementary graph (oh to be able to use colour!)

As a final example of a Honeycomb Lattice Labyrinth, the complementary graph for case (**10,1**) is shown below; I've left you the usual painstaking but satisfying job of revealing a central and six surrounding supertiles by drawing over unused lattice links to construct the tessellation graph. You may need to add more portions of the complementary graph round the edges.

As with Honeycomb (**9,3**), the complementary graph is an elegant composition of cartwheels, hexagons, triangles and trigonal spirals, and this time analogous constructions seem to work for all (**e,1**) Honeycomb Labyrinths. Photocopy the the pages inside the back cover to produce more blank lattices, or use the computer, and try another example, such as (**7,1**) or (**13,1**) As the value of **e** goes up (in threes) you will find that the left-or right-handedness of the spiral switches and that the numbers of hexagons and of triangles goes up in jumps. There's LOTS of scope for investigating rules for sub-families of Honeycomb Labyrinths AND you can cut each supertile into six to make a new family!

Honeycomb Labyrinth (10,1)  R = S = 222   - complementary graph

Instead of dismembering the Honeycombs, notice (see page 37) an intriguing property of the Honeycomb tessellation. Its complementary graph is itself a monohedral tessellation of diamond-shaped supertiles, each made up of two triangles. This tessellation displays the same **632** symmetry pattern as the Trefoil and Honeycomb Labyrinths. The illustration below shows how the symmetry axes are disposed, and how three shades can be used to distinguish the sets of diamonds lying in each of three orientations, at 1200 to each other.

It is natural to call this the Diamond Tessellation, which has a supertile area, **S**, of **2** triangles and a repeat unit, **R**, of **6** triangles. Is the Diamond tessellation the basic tessellation of a family of *Diamond Lattice Labyrinths*? These would need to display **632** symmetry, with all lattice points connected by the tessellation graph, and having the Diamond property of six diagonally-symmetrical supertiles meeting at each hexad axis and three at each triad axis, with a dyad axis on the lattice link at the centre of each supertile. It will be no surprise, by now, to learn that this family does exist.

The tessellation graphs of Honeycomb and Diamond Labyrinths are each other's complementary graphs, but the complementary graphs of higher-order Honeycomb Labyrinths we've drawn are not the tessellation graphs of monohedral tessellations, so we have to search further for the complementary graphs of higher-order Diamond Labyrinths.

It does not take long to find some examples of Diamond Lattice Labyrinths and to try to make sense of their complementary graphs. Have a look at the examples opposite. Some complementary graphs can be constructed from familiar elements, for instance those left for you to finish on pages 40 and 41, but others are hard to make sense of.

An alternative complementary graph is given for case **(4,1)**, generating what I call a beak-to-beak tessellation. The other **(4,1)** complementary graph generates a simpler shape, which can be characterised as standard for the **(a, a-3)** sub-family. Supertiles of this series of standard **(a, a-3)** Diamond Labyrinths are also shown opposite – we don't need complementary graphs to find and construct tessellations of these predictable shapes.

The supertiles of Diamond labyrinths must contain an even number of triangles, and there are three supertiles in the repeat unit, so Diamond Labyrinths must obey:

$$R = 3S = 2(e^2 + ef + f^2) = 6n \quad \text{(where } \mathbf{n} \text{ is a positive integer)}$$

This is the same Diophantine equation as for Honeycomb Labyrinths, so we expect Diamond Labyrinths to be drawable in all cases where $\mathbf{e} = \mathbf{f} + 3\mathbf{r}$, **where r is zero or a positive integer** (c.f. page 32). HOWEVER, attempts to set out the superlattice points and symmetry axes show that Diamond Labyrinths are only possible where at least one spacing parameter is odd. They CAN'T be drawn if **e** and **f** are both **even**, or one **even**, one **zero**. In such cases, the dyad axes fall not at lattice link centres but on lattice points.

Another family, the Dart Labyrinths can be drawn for (**even, even** or **zero**) separation parameters. In Dart Labyrinths four supertiles meet at each dyad axis, and six supertiles make up one repeat unit. Some low-order Diamond and Dart Labyrinths are shown below.

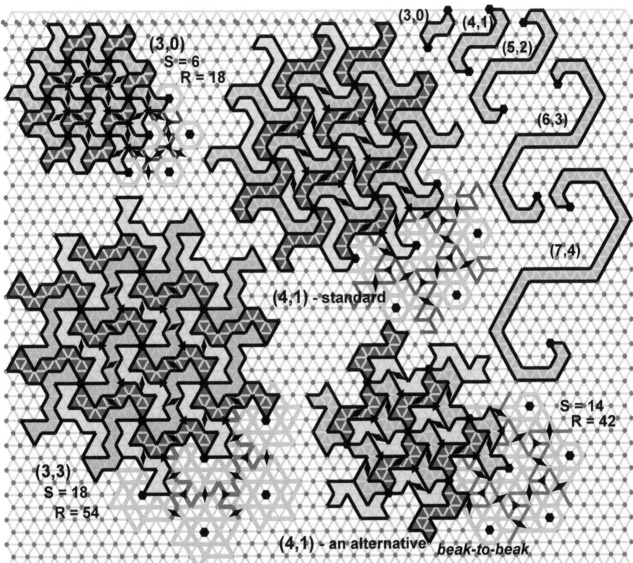

(3,0)

(4,1)

(5,2)

(6,3)

(7,4)

(3,0)
S = 6
R = 18

(4,1) - standard

(3,3)
S = 18
R = 54

S = 14
R = 42

(4,1) - an alternative *beak-to-beak*

**Some Diamond Labyrinths & complementary graphs + standard (e, e-3) supertiles**

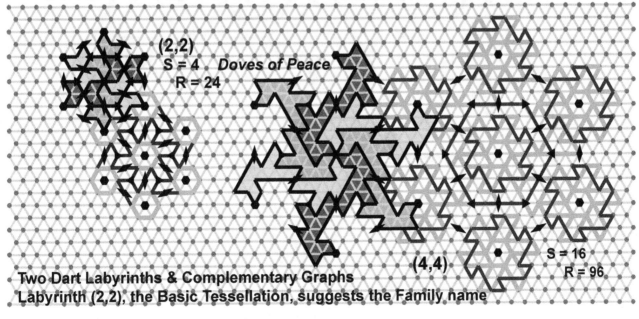

(2,2)
S = 4   *Doves of Peace*
R = 24

(4,4)

S = 16
R = 96

**Two Dart Labyrinths & Complementary Graphs**
**Labyrinth (2,2), the Basic Tessellation, suggests the Family name**

**Diamond labyrinth (7,1)**
**S = 38**
**R = 114**
*Running Postmen*

Complementary graph
of hexagons, triangles,
trigonal spirals and *spars.*

Dart Labyrinth (6,6)
*Climbers, Giftbearers?*
S = 36

R = 216

How DID I find this
complementary graph?

The similarity to Diamond (7,1)
is compelling but superficial.

# The Final Not-quite-so-small Print

First of all, a summary of the Lattice Labyrinth families we've laid out on the triangular lattice. We began with the Trefoil Labyrinths, followed by the Honeycomb Labyrinths. Gratifyingly, these two families use up all the possible sets of separation parameters (**e,f**).

We then found that the separation parameters possible for Honeycomb Labyrinths could be completely parcelled out between the Diamond and the Dart Labyrinths.

For the equations that determine who gets which set of parameters, see the sections devoted to each family. For those who don't like the algebra, we can set out all the possible separation parameters in a table which show, at rather more than a glance, which parameters will lead to which family of Labyrinths.

| 1,0 | 2,0 | 3,0 dia | 4,0 | 5,0 | 6,0 dar | 7,0 | 8,0 |
|---|---|---|---|---|---|---|---|
| 1,1 dia | 2,1 | 3,1 | 4,1 dia | 5,1 | 6,1 | 7,1 dia | 8,1 |
| | 2,2 dar | 3,2 | 4,2 | 5,2 dia | 6,2 | 7,2 | 8,2 dar |
| | | 3,3 dia | 4,3 | 5,3 | 6,3 dia | 7,3 | 8,3 |
| | | | 4,4 dar | 5,4 | 6,4 | 7,4 dia | 8,4 |
| | | | | 5,5 dia | 6,5 | 7,5 | 8,5 dia |

In the above table, the pairs of numbers represent the separation parameters which determine how far apart we lay the superlattice points on the lattice – hitherto I've usually referred to them as (**e,f**). In the table, I've left out the brackets. The pairs of separation parameters shown in normal type correspond to Trefoil Labyrinths, the pairs shown in bold type to Honeycomb Labyrinths. Of the separation parameters that give Honeycomb Labyrinths, those with **dia** appended correspond to Diamond Labyrinths, those with **dar** appended correspond to Dart Labyrinths.

This book is supposed to be both a source for artists and a work of mathematical recreation, so I do hope you have found it both accessible and enjoyable, though you may have felt the going getting harder and harder. Don't despair, this is perfectly normal. Books with mathematics in them may accelerate slowly, like trains pulling out of a station, but often go faster, and faster, and faster! I've too often given up the chase myself. Have another try.

The joy of mathematical recreations is that you start with some object or phenomenon that is quite everyday and particular – in this book it was the chessboard – and then find that you can look at this singular object as just one member of wide, often exotic classes of phenomena, exhibiting spectacular family relationships. In this book we have seen that the chessboard is but one member of a family of families of tessellations, each comprised of an infinite number of increasingly intricate members, many of which we can draw using the powerful complementary graph *algorithm* (a term honouring Arabian mathematics).

A further joy is finding examples of how geometry and algebra can each describe the same phenomena, and how geometry can make algebra visible and even "prove" algebraic expressions. In this book, Diophantine equations hold sway over geometric art, and the artwork conversely embodies expressions for the area of triangles and sums of squares.

Lastly, Lattice Labyrinth Tessellations appear to constitute an unexplored field of art and mathematics – surely rare these days. The scope for you to find new patterns is literally infinite and there must be new avenues to explore which my tramlines-mind has missed.

**Follow Dave's blog and contribute your own designs at www.latticelabyrinths.net**

Lightning Source UK Ltd.
Milton Keynes UK
UKOW07f1719180717

305575UK00017B/323/P